CREATURE FEATURES: CLASSIFY ANIMALS!

IT'S A REPTILE!

BY NATALIE HUMPHREY

Please visit our website, www.garethstevens.com. For a free color catalog of all our high-quality books, call toll free 1-800-542-2595 or fax 1-877-542-2596.

Cataloging-in-Publication Data
Names: Humphrey, Natalie.
Title: It's a reptile! / Natalie Humphrey.
Description: Buffalo, NY : Gareth Stevens Publishing, 2025. | Series: Creature features: classify animals! | Includes glossary and index.
Identifiers: ISBN 9781482466898 (pbk.) | ISBN 9781482466904 (library bound) | ISBN 9781482466911 (ebook)
Subjects: LCSH: Reptiles–Juvenile literature.

Published in 2025 by
Gareth Stevens Publishing
2544 Clinton Street
Buffalo, NY 14224

Designer: Andrea Davison-Bartolotta
Editor: Natalie Humphrey

Photo credits: Cover Kurit afshen/Shutterstock.com; p. 5 (top) Jan Bures/Shutterstock.com; pp. 5 (bottom left), 9 (top left) Kurit afshen/Shutterstock.com; p. 5 (bottom right) Fotojibb/Shutterstock.com; p. 7 (top) Joe McDonald/Shutterstock.com; p. 7 (bottom) Simlinger/Shutterstock.com; p. 9 (top right) iYodstocker/Shutterstock.com; p. 9 (bottom right) Super Prin/Shutterstock.com; p. 9 (bottom left) Nazzu/Shutterstock.com; p. 11 Shadska Photo/Shutterstock.com; p. 13 (inset) Landshark1/Shutterstock.com; p. 13 (main) Ralfa Padantya/Shutterstock.com; p. 15 seasoning_17/Shutterstock.com; p. 17 Breck P. Kent/Shutterstock.com; p. 19 (both) Marc Pletcher/Shutterstock.com; p. 21 (monitor) Juha Sompinmaeki/Shutterstock.com; p. 21 (sea snake) Rich Carey/Shutterstock.com; p. 21 (timber rattlesnake) Paul Staniszewski/Shutterstock.com.

Printed in the United States of America

CPSIA compliance information: Batch #CS25GS: For further information contact Gareth Stevens, New York, New York at 1-800-542-2595.

CONTENTS

Boldface words appear in the glossary.

Is That a Reptile?

Reptiles can be found around the world. The main reptile groups include turtles, lizards, snakes, and crocodiles and alligators. Some of these reptiles may even be living in your backyard! How do you know if you've spotted a reptile?

What Makes a Reptile?

Reptiles are **cold-blooded** animals. They have dry **scales** on their body. Some reptiles, such as alligators, have bony plates called scutes on their bodies as well as scales. All reptiles have tails. All reptiles have **lungs** to breathe air.

Scales

Some reptile scales, such as most snake scales, are smooth. Other reptile scales are somewhat bumpy. Scales can be colorful or help the reptile **camouflage** in their home. Many lizards and snakes can even change the color of their scales!

SNAKE

CROCODILE

Reptile Claws

Most reptiles, except for snakes and legless lizards, have claws. Reptiles use their claws to dig, catch food, and keep themselves safe from danger. Some reptiles, such as anoles, use their claws to climb.

CLAWS

Shedding Their Skin

Reptile scales don't grow with the reptile. Because of this, all reptiles **shed** their skin! Some reptiles, including many lizards and snakes, may shed their skin in one piece. Others, such as tortoises, shed small parts of their skin at a time.

Life in Eggs

Most reptiles start their lives in soft-shelled eggs. Many mother reptiles dig holes and bury their eggs to keep them safe from predators. Some reptiles build nests. Most reptiles leave their eggs, but some stay with their eggs until they **hatch**!

Live Babies

The eggs of some reptiles, such as garter snakes, stay inside the mother. The eggs hatch inside of her, and the babies are born live. Some baby snakes and lizards are never in eggs at all! Many boa constrictors have live babies.

Baby Reptiles

Most reptile parents don't care for their babies. A few reptiles, like crocodiles and American alligators, may stay with their young after they're born. Mother American alligators even carry their babies to the water after they hatch!

Reptiles Around the World

Most often, reptiles are found in warm places around the world. A few reptiles, such as the timber rattlesnake, can be found in places that get colder. Other reptiles, like **desert** monitors, live in deserts. Sea snakes live in the ocean!

GLOSSARY

camouflage: Colors or shapes on animals that allow them to blend in with their surroundings.

cold-blooded: Having a body temperature that's the same as the temperature of the surroundings.

desert: An area of land that gets very little rain.

hatch: To break open or come out of.

lung: The part of an animal that takes in air when it breathes.

scale: One of the flat plates that cover a reptile's body.

shed: To lose fur or skin.

FOR MORE INFORMATION

BOOKS

Barder, Gemma. *Be a Reptile Expert.* New York, NY: Crabtree, 2025.

Lowe, Lindsey. *Reptiles.* Tucson, AZ: Brown Bear Books, 2023.

WEBSITES

Britannica Kids: Reptile
kids.britannica.com/kids/article/reptile/353708
Learn more about what makes a reptile a reptile.

National Geographic Kids: Reptiles
kids.nationalgeographic.com/animals/reptiles
Discover more reptiles and find out how they live in the wild.

Publisher's note to educators and parents: Our editors have carefully reviewed these websites to ensure that they are suitable for students. Many websites change frequently, however, and we cannot guarantee that a site's future contents will continue to meet our high standards of quality and educational value. Be advised that students should be closely supervised whenever they access the internet.

INDEX